# *Meet me at*™
# *The* BELLE MEADE PLANTATION

# Meet me at™

# The BELLE MEADE PLANTATION

## TIMELESS IMAGES AND FLAVORFUL RECIPES
## FROM THE QUEEN OF TENNESSEE PLANTATIONS

**Andrew B. Miller**
*Executive Editor*

**Mary Lawson**
*Editor*

**Daisy King**
*Recipe Editor*

Historic
HOSPITALITY

*Meet me at the Belle Meade Plantation* was published by Historic Hospitality in collaboration with Belle Meade. Historic Hospitality creates exquisitely designed custom books for America's iconic hotels, inns, resorts, spas, and historic destinations. Historic Hospitality is an imprint of Southwestern Publishing Group, Inc., 2451 Atrium Way, Nashville, TN 37214. Southwestern Publishing Group is a wholly owned subsidiary of Southwestern/Great American, Inc., Nashville, TN.

**Christopher G. Capen,** *President,* Southwestern Publishing Group, Inc.
**Sheila Thomas,** *President and Publisher,* Historic Hospitality
**Andrew B. Miller,** *Executive Editor*
**Mary Lawson,** *Editor* | **Joey McNair,** *Page Designer*
**Daisy King,** *Recipe Editor* | **LeAnna Massingille,** *Cover Designer*
www.historichospitalitybooks.com | 800-358-0560

*Special thanks to Alton Kelly, John Lamb, Jenny Lamb, Mary Bray Wheeler,*
*and Ridley Wills II for their assistance in the development of this book.*

ISBN: 978-0-87197-627-7
Printed in China
10 9 8 7 6 5 4 3

# The BELLE MEADE PLANTATION

*The Belle Meade Mansion stands proudly as one of the South's most outstanding showplaces. Named after its surroundings, French for "beautiful meadow," its fascinating history includes the grandeur of world-renowned fame.*

# JOHN HARDING
# SETTLES IN TENNESSEE

In 1807, Virginian John Harding bought Dunham's Station log cabin and 250 acres near the Natchez Trace seven miles from Nashville, Tennessee. He began boarding horses for neighbors such as Andrew Jackson and advertising in Nashville newspapers for horses standing stud at his farm. By 1816, he was breeding thoroughbreds and soon racing his own horses locally. John registered his own Racing Silks with the Nashville Jockey Club in 1823 and was training horses on the track at his McSpadden's Bend Farm. As horse racing became more popular in the South, he had made enough money to build a new Federal-style brick home and eventually increased the size of his plantation to twelve hundred acres. During his three decades of management, Harding sold blacksmith services, dressed lumber, and sold farm products, shipping grain as far as Charleston and New Orleans. It was the horses, however, that eventually catapulted Belle Meade into fame and fortune.

JOHN HARDING.

# WILLIAM HARDING
# AND BELLE MEADE'S DESTINY

John's son, William Giles Harding, lived on the McSpadden's Bend property and worked with his father training horses. By the time William Giles, a college-educated general in the state militia, assumed management of the Belle Meade plantation in 1839, he was keenly interested in all aspects of breeding and racing. By 1842, the first famous stud at Belle Meade, Priam, had achieved the title of American Thoroughbred Horse of the Year and again in 1844-1846. Belle Meade's historic journey into the realm of world-class horses had begun, and by the 1850s, William was one of Tennessee's wealthiest men and one of the larger landowners. He began enlarging his home, doubling its size and transforming it into a showplace. In 1853, he added the front porch and massive limestone columns to his mansion, creating a stunning example of the grandeur of the South's Greek Revival Antebellum architecture.

# ELIZABETH MCGAVOCK CONFRONTS THE CIVIL WAR

When the dark clouds of the Civil War loomed in the distance, General Harding was appointed to a three-man committee charged with the responsibilities of equipping an army for the South. He devoted most of his time to the cause and relied on his wife, Elizabeth McGavock, and his overseer to manage the plantation. He donated $500,000 to the Southern cause and Andrew Johnson, military governor of Tennessee, decided to use Harding as an example to other Confederate sympathizers by exiling him to prison in Fort Mackinac, Michigan. Tirelessly assuming her husband's responsibilities during his incarceration, McGavock was left to manage the plantation and look after her "family of 150 persons," mostly slaves. Primarily through her heroic efforts, Belle Meade survived the Civil War. At one point in 1863, she was awakened by the commotion of drunken Confederate soldiers and horses on her front lawn and hastily grabbed a loaded pistol as she ran down the stairs. Aiming the pistol directly at them, she demanded that they leave the premises at once. Stunned and in disbelief, the leader of the group turned and called out, "Mount and get away, boys. We don't want to disturb no lady as game as that!"

# WAR SURROUNDS BELLE MEADE

The Civil War eventually arrived on the doorsteps of Belle Meade when Confederate Gen. James R. Chalmers of Nathan Bedford Forrest's cavalry set up his headquarters prior to the Battle of Nashville in 1864. On the first day of the battle, Union soldiers burned the Rebel wagons parked at the racetrack while Chalmers was elsewhere. Returning to Belle Meade, Chalmers's men charged the Yankees and drove them back before running into an enemy infantry camp. The Yankees fired as the cavalry galloped back past the mansion, where nineteen-year-old Selene Harding waved a handkerchief despite the bullets flying around her. The scars from the bullets are still visible in the porch columns today. The Civil War almost destroyed horse breeding in Tennessee, due to secession and massive troop movements throughout the state. Though it had interrupted breeding and racing in the South, General Harding was able to keep all of his thoroughbreds, even while other farms were having their horses requisitioned by both armies.

# BELLE MEADE
# AND ITS LEGACY

In 1867–1868, General Harding won more purses with his own horses than any man living at that time in the United States. In 1867, he held the first sale of horses bred on his farm. He was the first in Tennessee to use the auction system for selling thoroughbreds, and Belle Meade became the most successful thoroughbred-breeding farm the state would ever see. Other famous horses at the plantation arrived after the Civil War. William Giles Harding acquired Bonnie Scotland in 1872, making Belle Meade a world famous stud farm. He was the leading American sire in 1880 and 1882, and many of his descendants are still racing today. In recent years, almost all of the horses entered into the Kentucky Derby, as well as in other major stakes races, can be traced back to the Belle Meade Plantation and are descendants of the great sire Bonnie Scotland.

# WORKING TOGETHER

After the war, William Harding gave control of the farm to his son-in-law, William Hicks Jackson, a West Point graduate who had commanded a cavalry division under General S. D. Lee in Mississippi and Louisiana. General Jackson, also an avid horseman, began working with his father-in-law to expand the breeding farm, and by 1875, they had decided to retire the racing silks and concentrate on breeding. The legacy of Belle Meade as a foundation for America's thoroughbred horse lineage is credited as much to the African American trainers and jockeys who worked the farm as to the Harding family who owned the property. The Kentucky Derby at Churchill Downs in Louisville has been run annually since 1875. The first winner was Aristides, who earned $2,850 for the win. The horse was trained by an African American man, Ansel Williamson, and raced by African American jockey, Oliver Lewis, one of fifteen black jockeys who rode in the Derby.

# GENERAL JACKSON MODERNIZES BELLE MEADE

After General Harding's death, General William Hicks Jackson assumed one-third ownership of the horse farm with his wife's half-brother John and General Jackson's brother Howell. General Jackson was, however, the only co-owner working as daily manager, and he greatly expanded the home. By 1883, he was modernizing the interior, including adding three full bathrooms complete with hot and cold running water, an updated kitchen, and a telephone. Under his tutelage, the vast estate became a prosperous 5400-acre plantation housing a 600-acre deer park, and selling breeding stock for Alderney cattle, Cotswold sheep, and Cashmere goats. In the meantime, Belle Meade remained in the world's spotlight, as Iroquois, acquired by Billy Jackson in 1886, became the first American owned and bred horse to win the English Derby in 1881.

# BELLE MEADE GAINS WORLDWIDE ATTENTION

The prominent family hosted many dignitaries such as General U. S. Grant, General William Sherman, and Robert Todd Lincoln, but it was the visit of President Grover Cleveland and his wife, along with the purchase of Iroquois, the first American horse to win the English Derby, that put Belle Meade on the map. It had now achieved world recognition as a thoroughbred horse farm and a host to presidents. At a time in history when sports heroes were horses, the farm was serving as a thoroughbred nursery, famous for the breeding and training of racehorses. Its reputation as a first-class breeding establishment attracted buyers from around the world for the annual yearling sales. Amazingly, most of the Kentucky Derby winners of the twentieth century can trace their lineage to Bonnie Scotland, the greatest sire of all time, who was a product of Belle Meade. Among his descendants are Seabiscuit, Seattle Slew, Affirmed, and Secretariat. The bloodlines for more recent Kentucky Derby winners, Funny Cide and Barbarro, can also be traced back to Belle Meade.

# BELLE MEADE SOLD AT AUCTION

The deaths of William Hicks Jackson and his son William Harding Jackson, both in 1903, ushered in the final days of Belle Meade Plantation. Afterward, the ownership of Belle Meade Plantation passed to William Harding Jackson's sister, Selene Elliston, his two-year-old son, William Harding Jackson Jr,. and the Catholic Church. The new century had also brought on financial hardship with the advent of anti-gambling laws. Poor spending habits and a weakened economy had led the family into serious debt. Following Jackson's death, at a time when Belle Meade was the oldest and largest thoroughbred farm in the United States, the vast estate was sold at auction. By 1906, all of the 2600 acres that at one time belonged to Selene and General Jackson were gone and the fourth generation of the Harding family moved off the property. The Belle Meade Land Company, a business syndicate, consolidated the mansion and acreage for future development. On May 19, 1906, John Overton Dickinson, a second cousin to Elizabeth Harding, purchased the Belle Meade Mansion and forty surrounding acres. Dickinson moved into the mansion with his family.

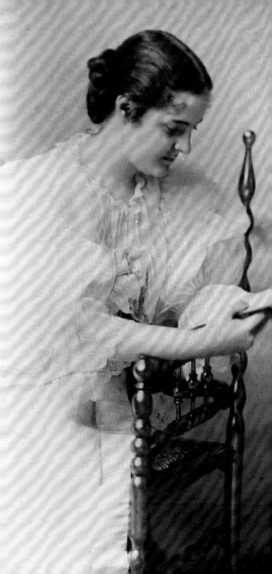

# BELLE MEADE RESTORED

In 1953, the Belle Meade Mansion and eight outbuildings on thirty acres were deeded to the Association for the Preservation of Tennessee Antiquities, and are managed today by the Nashville chapter of the Association. The former plantation lands now form the independent city of Belle Meade, Tennessee. In 1986, Irene Wills, a descendant of the Jackson family, developed a restoration committee for raising funds to restore the Belle Meade house to its late nineteenth century appearance. Utilizing a series of photographs commissioned by the Jackson family in 1900, two images of the interior of the house documented original carpet patterns, furniture placement, and original light fixtures. During the restoration, the reproduction wallpaper was removed, and the walls and ceiling were painted gray as they were in 1883. Furnishings were returned to their nineteenth century locations and the room looks today as it did at the turn of the century. Many of the objects in the 1900 image of the main hall are in the Belle Meade collection, including the thoroughbred paintings on the walls and the empire style sofa. Today, the Belle Meade mansion is an elegant home and appears just as it did in the late nineteenth century.

# THE OLD SOUTH RELIVED

As a result, the Belle Meade Mansion still remains a national treasure and one of Nashville's most popular attractions. *Budget Travel Magazine* has named it one of "America's Grandest Mansions." A visit to Belle Meade allows one to leave the hustle of the world behind and step back into the elegance of the antebellum South. Now a museum, it is a storyteller's dream. A tour includes costumed guides who share fascinating glimpses into the lives of its colorful residents and their notable guests. The magnificent entry hall is adorned with the paintings of its world famous horses. Its stately rooms are filled with original furnishings and period antiques, bringing to life its southern charm.

# BELLE MEADE TODAY

A tour across the idyllic grounds includes the 1832 slave cabin containing a permanent exhibit documenting the contribution of African Americans, a colossal carriage house containing an extensive antique carriage collection, and stables which once housed carriage and riding horses that were owned by the Jackson family. The original log cabin, built in 1790, is one of the oldest houses in Tennessee. The impeccably landscaped gardens and other outbuildings, including a smokehouse and dairy, give a glimpse of the practical and pastoral side of the Old South. Added to the National Register of Historic Places in 1969, Belle Meade now consists of a more manageable thirty acres. It abounds with the activity of tour buses, music events, festivals, art exhibits, educational programs, summer camps, and hosts weddings and corporate events. Belle Meade is still indeed, "The Queen of Tennessee Plantations."

# A Civil War Reunion

General William Giles Harding was a great supporter of the Confederates at the dawn of the Civil War. Harding did not fight in the war, but did what he could to contribute to the Confederate cause. Nashville newspapers reported that he gave $500,000 to support them, including giving money to arm and equip an artillery unit, The Harding Light Artillery, which was named in his honor. In September of 1884, the general invited the unit's survivors and their wives or girlfriends to attend a reunion and barbeque picnic on the front lawn of Belle Meade. The men played a game of baseball and everyone enjoyed music by the Enterprise Cornet Band of Nashville. Later, the men posed in front of the house with the family for a photograph before departing on the evening train.

# SALT RISING BREAD

## YEAST

4 small Irish potatoes (shred)
3 tablespoons sugar
4 tablespoons corn meal

1 teaspoon salt
1 quart boiling water poured over the above

Mix well. Cover and keep in a warm place over night.

Sponge.

Strain yeast and add:

1 cup meal
2 cups warm water or milk

1/4 teaspoon baking soda
flour enough to make stiff batter

Mix well, place in bowl, and set in pan of fairly hot water. Cover. The sponge should rise in an hour. Keep water one inch above where sponge comes to in its container.

## DOUGH

1 cup Crisco
2 teaspoons salt

1 tablespoon sugar

After sponge has risen, add enough flour to make bread consistency. Knead well, but do not let it get chilled. Make into loaves and put into well greased pans. Let rise until double its size. Bake in a medium oven for an hour.

*Yields 1 loaf.*

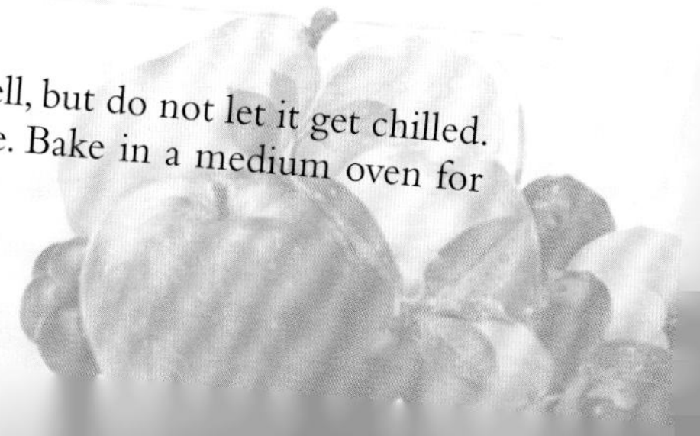

# MISS DAISY'S BACON-MARMALADE ROUNDS

| | |
|---|---|
| 1 | pound sharp cheddar cheese, grated and softened |
| 8 | ounces cream cheese, softened |
| 3 | egg yolks, lightly beaten |
| 1/2 | cup orange marmalade |
| 36 | Melba rounds (crackers) |
| 1/2 | pound cooked bacon, drained, and crumbled |

Assemble all ingredients and utensils.

In a large bowl of an electric mixer, combine chedder cheese, cream cheese, and egg yolks.

Blend in marmalade. Spread the mixture on Melba rounds. Top with bacon. Broil until hot and bubbly.

*Yields 3 dozen rounds.*

# BROCCOLI WITH GORGONZOLA DIP

| | |
|---|---|
| 3 to 4 | pounds blanched, chilled broccoli flowerets |

## DIP

| | |
|---|---|
| 3 | cups crumbled white Gorgonzola cheese |
| 3 | tablespoons canola oil |
| 3 | tablespoons lemon juice |
| 1 | tablespoon lemon zest |
| 3 | eggs, at room temperature |
| 3 | egg yolks, at room temperature |
| 1/2 | teaspoon salt |
| | Crisp-cooked crumbled bacon for garnish |

Assemble all ingredients and utensils.

Arrange broccoli on a large serving platter, leaving room in the center for the bowl of dip. Just before serving, combine cheese, oil, lemon juice, and zest in a blender container. Process to paste consistency.

Add eggs, egg yolks, and salt. Pulse to blend, about 1 minute. Turn the dip into a serving bowl and sprinkle with crumbled bacon. Serve immediately.

*Yields 20 to 25 servings (approximately 3 1/2 cups dip).*

# PECAN CHEESE WAFERS

| | | |
|---|---|---|
| 8 | ounces sharp cheddar cheese | 1 cup finely chopped pecans |
| 1 | cup butter, softened | 1/8 teaspoon cayenne pepper |
| 1 1/2 | cups self-rising flour | |

Assemble all ingredients and utensils. Grate cheese. Let it soften with butter in a bowl. Mix in remaining ingredients. Roll dough into rolls about 1 1/2 inches diameter. Place on waxed paper. Chill for several hours. Slice in thin wafers about 1/8 inch thick. Bake in a 350-degree oven about 8 minutes or until browned.

*Yields about 75 wafers.*

To Mrs. Jackson—
whom I have never
seen, but
Thos. Lee Hasbord.

# FARM PUMPKIN BREAD

| | |
|---|---|
| 1 cup water | 3 1/2 cups self-rising flour |
| 1 cup vegetable oil | 1 teaspoon each: salt, |
| 1 pound can pumpkin or | ginger, and nutmeg |
| pumpkin pie filling | 1/2 teaspoon baking powder |
| 3 cups sugar | 1/2 teaspoon cloves |
| 3 eggs | 2 teaspoons cinnamon |
| 1 cup black walnuts, chopped | 2 teaspoons baking soda |
| 1 1/2 cups dates, chopped | |

Assemble all ingredients and utensils. In a large mixing bowl, combine first seven ingredients. Sift together remaining ingredients. Mix well with pumpkin mixture. Pour into 2 large greased loaf pans. Bake in a 325-degree oven for 1 and 1/2 hours.

*Yields 2 large loaves.*

# PRESIDENT CLEVELAND VISITS BELLE MEADE

After General Jackson purchased Iroquois in 1886, many prominent men were drawn to the Belle Meade farm. In 1887, the Jackson family hosted President Grover Cleveland and his new wife. The entire household bustled as they prepared for the distinguished guests. General Jackson entertained the local press in his study, the staff hung Japanese lanterns in the trees lining the main drive as a carriage with liveried servants waited at the Belle Meade depot for the arrival of the train.

Upon their arrival, the president and his entourage were escorted up the drive to the main house where the Jackson family waited on the porch to greet them. The guests spent the night in an elegant suite of bedrooms and enjoyed breakfast with the family the next morning. Following breakfast, the group took a walk in the deer park where grooms mounted on their horses rushed a herd of deer past President and Mrs. Cleveland. The couple departed in the afternoon to visit Mrs. James K. Polk before leaving the city. This presidential appearance at the Belle Meade farm was featured in many newspapers, providing it with publicity that served to increase Jackson's business and success on a national level.

## CORN LIGHT BREAD

| | |
|---|---|
| 2 | cups plain corn meal |
| 3/4 | cup sugar |
| 1/2 | cup flour |
| 1/4 | teaspoon baking soda |
| 1 | teaspoon salt |
| 2 | cups buttermilk |
| 3 | tablespoons shortening, melted |

Assemble all ingredients and utensils. In a bowl, mix dry ingredients with buttermilk and shortening. Pour into a greased loaf pan. Bake in a 350-degree oven for 45 minutes or until done. Turn out on a rack and cool.

*Yields 1 loaf, 8 to 12 servings.*

## BEATEN BISCUITS

| | |
|---|---|
| 3 | cups sifted all-purpose flour |
| 1/2 | teaspoon sugar |
| 1/2 | teaspoon salt |
| 3 | tablespoons cold butter |
| 3 | tablespoons cold lard or vegetable shortening |
| 1/2 | cup cold milk |
| 1/2 | cup cold water |

Assemble all ingredients and utensils. Sift flour, sugar, and salt into a bowl. Add butter and lard; blend with pastry blender or forks until mixture looks like coarse corn meal. Add milk and water, tossing mixture with a fork. Knead 15 minutes; then beat 15 minutes or less with a rolling pin. Cut with small floured biscuit cutter. Prick tops 3 times with a fork. Place on baking pan. Bake in a 325-degree oven for 25 to 30 minutes. Biscuits will be a very light brown. Serve cold.

*Yields about 34 to 36 biscuits.*

# CELEBRATION WASSAIL

5 oranges with whole cloves
2 cups water
5 cups pineapple juice
3 quarts apple cider
2 sticks cinnamon

1/2 cup honey
1/2 teaspoon nutmeg
1/3 cup lemon juice
2 teaspoons lemon rind
2 cups rum

Assemble all ingredients and utensils. Stud 5 oranges with whole cloves, about 1/2 inch part. Place in baking pan with 2 cups water. Bake in a 350–degree oven for 20 minutes. Meanwhile, heat cider and cinnamon sticks and honey in a large pot. Bring to boil over medium heat; simmer covered for 5 minutes. Add remaining ingredients. Pour over spiced oranges that have been transferred from oven to punch bowl.

*Yields about 40, 4-ounce cups.*

# ANGEL BISCUITS

| | | | |
|---|---|---|---|
| 5 | cups all-purpose flour | 3/4 | cup butter |
| 4 | tablespoons sugar | 1 | envelope dry yeast |
| 1 | tablespoon baking powder | 3 | tablespoons warm water |
| 1 | teaspoon salt | 2 | cups butter milk |
| 1 | teaspoon baking soda | 1/8 | cup melted butter |

Assemble all ingredients and utensils. In a deep bowl, sift dry ingredients together; cut in butter with fork or pastry blender until mixture is crumbly. Dissolve yeast in water. Stir yeast and buttermilk into flour mixture. Roll out onto a floured board about 1/2-inch thickness. Cut with biscuit cutter. Brush with melted butter. Place on a baking sheet and bake in a 400-degree oven 10 to 12 minutes or until brown.

*Yields 24 to 36 biscuits.*

# COFFEE PUNCH

| | |
|---|---|
| 2 quarts strong brewed coffee (8 cups) | 1/2 cup sugar |
| 1 pint cold milk (2 cups) | 2 quarts vanilla ice cream |
| 2 teaspoons vanilla extract | 1/2 pint heavy cream |
| | 1/2 tablespoon ground nutmeg |

Assemble all ingredients and utensils. In a deep bowl, combine coffee, milk, vanilla, and sugar. Chill. Break ice cream into chunks in punch bowl just before serving; pour chilled coffee mixture over ice cream. Whip cream, spoon into mounds on top of punch. Sprinkle with nutmeg.

*Yields 18 servings.*

# IROQUOIS COMMANDS WORLDWIDE FAME

Iroquois was the first American-bred thoroughbred racehorse to win the prestigious English Derby in 1881. Millionaire Pierre Lorillard IV of the tobacco and snuff family fame had purchased him in 1879, and though he never stood higher than fifteen hands two-and-a-half inches, Iroquois won four of his two-year-old races on British soil. When the Americans learned of Iroquois's English Derby win, they temporarily closed business on Wall Street to celebrate the victory. In June of 1881, England's legendary jockey, Fred Archer, rode Iroquois to a neck-and-neck win at the Epsom Derby. In September of that same year, Iroquois and Archer won against a field of fourteen in the St. Leger race. That year, Iroquois also won the Prince of Wales' Stakes and became the first horse, American or British, to win all three races in one year.

Iroquois was acquired by Billy Jackson in 1886 for $20,000 in gold and after arriving at Belle Meade, he was treated as the family pet. By the mid-1890s, at a time when the average stud fee was $100 to $300, Iroquois was advertised as a private stud commanding $2,500 for his services.

THE DERBY. One of the fleetest horses on the turf, distinguished as the first American horse to win the celebrated Derby races from the best English horses. Sired by "Leamington;" dam, "Maggie B. B." Owner, P. LORILLARD, New York.

IROQUOIS.

IROQUOIS

# BELLE MEADE PLANTATION SYLLABUB

| | | | |
|---|---|---|---|
| 3 | cups apple cider | 1 | cup sugar |
| 1/4 | cup lemon juice or Madeira wine | 2 | egg whites |
| 1 | teaspoon light corn syrup | 1/4 | cup sugar |
| 3 | tablespoons grated lemon rind | 2 | cups whole milk |
| | | 1 | cup light cream |

Assemble all ingredients and utensils. Mix all ingredients in a large bowl. Stir until sugar dissolves. Then refrigerate until cold. Just before serving, beat 2 egg whites until frothy; add 1/4 cup sugar, a small amount at a time until meringue stands in peaks. Then beat 2 cups milk and 1 cup of light cream into cider mixture. Pour into punch bowl and spoon meringue on top.

Garnish with nutmeg.

*Yields 12 servings.*

# APRICOT SALAD WITH CUSTARD

4 cups hot water
1 6-ounce package apricot gelatin
1 cup miniature marshmallows
1 14-ounce can crushed pineapple, drained, reserving juice
2 large bananas, mashed

1/2 cup sugar
2 tablespoons all-purpose flour
1 tablespoon butter
1 8-ounce package cream cheese, softened
1 4-ounce carton whipped topping

Assemble all ingredients and utensils. In saucepan, heat water to boiling; add gelatin and marshmallows and stir until dissolved. Let cool. Add crushed pineapple and bananas. Pour mixture into a 13x9–inch glass casserole dish and refrigerate until salad congeals.

In a saucepan combine reserved pineapple juice, sugar, flour, and butter. Cook until thick, add cream cheese and let cool. Fold in whipped topping and spread over top of congealed gelatin.

*Yields 12 servings.*

# WILTED SPINACH AND ORANGE SALAD

1 pound fresh spinach, washed and dried
8 green onions, chopped
3 oranges, peeled, cut into bite-size pieces
8 slices bacon
3 tablespoons bacon drippings
3 tablespoons vegetable oil
3 tablespoons lemon juice
1 tablespoon sugar
1/2 teaspoon salt

Assemble all ingredients and utensils. Stem spinach and tear leaves into bite-sized pieces. Toss with green onions in a salad bowl. Fry bacon in a skillet, reserve drippings. Crumble and set aside. Mix 3 tablespoons drippings, lemon juice, oil, sugar, and salt in a saucepan. Bring to a boil and pour over spinach and green onions. Add orange pieces and bacon.

Toss again and serve immediately.

*Yields 6 to 8 servings.*

# MISS DAISY'S TOMATO ASPIC

4 envelopes unflavored gelatin
1/2 cup water
4 cups tomato juice
3 tablespoons lemon juice
2 teaspoons salt
1 teaspoon onion juice
1/8 teaspoon cayenne pepper
1/4 cup sliced stuffed olives
4 stalks celery, chopped finely

Assemble all ingredients and utensils. Let gelatin stand in 1/2 cup water for 10 minutes. In a double boiler, add tomato juice, lemon juice, salt, onion juice, and cayenne pepper; bring to a boil. Add olives and celery. Pour into 1 1/2-quart mold or an 8x8-inch pan.

Refrigerate until congealed.

*Yields 8 servings.*

# BELLE MEADE STUD.

"Home of the Race-horse" is a title that has well been earned by Belle Meade."—[New York Sun.

"In many respects Belle Meade is the most remarkable breeding establishment in the world."—[Philadelphia Record.

"The oldest in years, Belle Meade has always kept in the front rank as America's greatest thoroughbred nursery.—[Chicago Tribune.

| STALLIONS IN USE SEASON OF 1892 | SERVICE FEE. |
| --- | --- |
| Iroquois | $300 |
| Luke Blackburn | $300 |
| Tremont | $300 |
| Inspector B. | $150 |
| Imp. Loyalist | $150 |
| Enquirer | Private |
| Imp. Great Tom | Private |

Ten approved mares each the limit. Return privilege to same horse.

Catalogue of stallions and broodmares mailed free on application after February 1, 1892.
ONE HUNDRED AND TWENTY-FIVE HORSES by Belle Meade sires won over 450 RACES on the American turf in 1891, aggregating over $300,000.
The Belle Meade yearlings will be sold at Tattersall's Sale Repository, Seventh Avenue and Fifty-fifth Street, New York City, on Monday, June 20th (Evening).

**W. H. JACKSON, Proprietor,**
P. O. Box, 383 Nashville, Tenn.

## STALLIONS
IN USE SEASON OF 1892
—AT THE—
ELMENDORF STUD,
LEXINGTON, KY

**FOXHALL.**

---

# ASHLAND!

## DICTATOR 113,

Sire of Jay Eye See 2:10, Phallas 2:13¾, Director 2:17 and 33 trotters in the 2:30 list and 3 pacers in the 2:25 list. Grandsire of Direct (p) 2:06; Nancy Hanks 2:09; Margaret S. 2:12½; Lockhart 2:14¾. Private Stallion.

## KING RENE 1278,

Record 2:30¼.

Sire of Fugue 2:19¼, Keeler 2:19½, Prince Edward 2:20, Sarcenet 2:20½, and of 17 trotters in the 2:30 list. Grandsire of Belle Archer (4) 2:15½. The captor of 18 herd premiums, without a single defeat; a record never approached, and the best evidence of the fact that ... breeds him for beauty, style ... Stands at $100.

---

## 1892.
NOW LOCATED AT
# KENMORE,
Lexington, Ky.

# VASCO,

LUKE BLACKBURN, 1877

"MAN'S NOBLEST FRIEND"
ERECTED TO THE MEMORY OF
ENQUIRER,
GREATEST OF RACE HORSE SIRES.
BY JOHN R. McLEAN PUBL'R
CINCINNATI ENQUIRER,
AUG. 26, 1897.

ENQUIRER

J. F. DELURY,   FINE TAILORING,
                    LOW PRICES,
LIVERIES,
                    RIDING HABITS.

Iroquois Winning the Derby

# MARINATED ASPARAGUS SALAD

4   15-ounce cans green
    asparagus spears
4   15-ounce cans white
    asparagus spears

1   bunch watercress
2   heads Boston lettuce

## LEMON VINAIGRETTE

1   8-ounce bottle herb-garlic
    salad dressing
1/2 cup lemon juice

2   tablespoons chopped chives
1/4 teaspoon white pepper

Assemble all ingredients and utensils. Drain asparagus spears. Prepare the vinaigrette by combining all ingredients in a small bowl. In a shallow dish, layer asparagus and pour dressing over. Refrigerate for several hours. To serve, arrange clean, crisp lettuce leaves and watercress on a serving platter. Alternate marinated asparagus (green and white) in sections. Place watercress in the center. Spoon the vinaigrette over all.

*Yields 25 servings.*

Uncle Bob Green 78 yrs.
Bellemeade.

# A MASTER HORSEMAN

Bob Green became famous for his working knowledge of horse flesh. His skill and experience as a hostler earned him one of the highest salaries ever paid to a horse hand of the day. Brought to work at the mansion in 1839 as a slave by General Harding, he grew up working with the horses and became an expert in everything related to the thoroughbred. It has been said that many a gentleman in the horse business owed a debt of gratitude to Green for his knowledge at the yearling sales. As head groom at Belle Meade, he always wore a white apron. Green was introduced to President Grover Cleveland during his visit in 1887 and led President Cleveland on a tour of the studs, including Iroquois, Bramble, Enquirer, and Luke Blackburn. Sadly, at the end of his life, Green was forced to move from the plantation and his home at the old family cabin. Not only had he been in charge of all thoroughbreds at Belle Meade, he owned and raced thoroughbreds during his lifetime and was heartbroken to leave Belle Meade. In 1906, he was granted his request for burial at the farm, where he rests today in an unmarked grave.

# CHILLED BROCCOLI SALAD

8 slices bacon, crisp-fried and crumbled florets and tender stem portions of 1 bunch broccoli, chopped

1/2 cup each finely chopped red onion and celery

1/2 cup raisins

1/2 to

3/4 cup cashews or pecans

1 cup mayonnaise

1/4 cup sugar

3 tablespoons red wine vinegar

Assemble all ingredients and utensils. Combine the bacon with the broccoli, onion, celery, raisins, and cashews in a large bowl and mix well.

Mix the mayonnaise, sugar, and vinegar in a small bowl. Add to the broccoli mixture and mix well. Chill, covered, for one hour or longer before serving. Garnish with additional bacon.

*Yields 6 to 8 servings.*

# TOMATOES FLORENTINE

2    10-ounce packages frozen chopped
     spinach, cooked and well-drained
1/2  green pepper, chopped
1/4  cup chopped onion
1    cup chopped celery
2    hard-boiled eggs, chopped
     salt and pepper to taste
1/4  cup mayonnaise
1    teaspoon lemon juice
6    tomatoes, hollowed and drained

Assemble all ingredients and utensils. Mix spinach with pepper, onion, celery, and eggs. Season; stir in mayonnaise and lemon juice. Spoon into tomato shells. Serve warm or cold.

*Yields 6 servings.*

# CORN PUDDING

2 1/2  cups cream-style corn
5      tablespoons all-purpose flour
1      tablespoon sugar
1      teaspoon salt
1/4    cup butter, melted
3/4    cup milk
3      eggs, beaten

Assemble all ingredients and utensils. Mix corn and flour with a wire whisk. Add remaining ingredients and mix well. Bake in a greased 2-quart casserole dish in a 325–degree oven for 1 hour.

*Yields 4 to 6 servings.*

# FRIED OKRA

| | |
|---|---|
| 1 | pound okra |
| 1/2 | cup corn meal |
| 1/2 | teaspoon salt |
| 1/4 | teaspoon pepper |
| 2 to 3 | tablespoons vegetable oil |

Assemble all ingredients and utensils. Slice okra into 1/2-inch rounds. In a paper bag, combine corn meal, salt and pepper; add okra and shake. Fry in hot oil until okra is browned. Drain on paper towels. Season with salt and pepper before serving.

*Yields 4 servings.*

# PARTY SQUASH

| | |
|---|---|
| 1 | pound yellow squash, sliced |
| 1 | teaspoon sugar |
| 1/2 | cup mayonnaise |
| 1/2 | cup minced onion |
| 1/4 | cup finely chopped green pepper |
| 1/2 | cup chopped pecans |
| 1 | egg, slightly beaten |
| 1/2 | cup grated cheddar cheese |
| | salt and pepper to taste |
| | bread or cracker crumbs |
| 1/4 | cup butter |

Assemble all ingredients and utensils. Cook squash, drain, and mash. Add remaining ingredients except crumbs and butter. Pour into a 2-quart casserole. Top with crumbs; dot with butter. Bake in a 350-degree oven for 35 to 40 minutes.

*Yields 6 servings.*

# VEGETABLE CASSEROLE

| | | | | |
|---|---|---|---|---|
| 1 | small cauliflower | 4 | tablespoons butter |
| 8 | small new potatoes | 4 | tablespoons all-purpose flour |
| 8 | small carrots | 2 | cups milk |
| 10 | small onions | 1 | teaspoon salt |
| 1 | cup canned or | 1 | teaspoon pepper |
| | frozen green peas | 1/2 | pound sharp cheese, grated |

Assemble all ingredients and utensils. Separate cauliflower into flowerets. Add next 3 ingredients; cook in water until tender. Drain well. Add drained peas. Place in a 2-quart casserole. Melt butter in a saucepan, add flour, cook until smooth, add milk, salt and pepper, and cook until thickened, stirring constantly. Add cheese, stir until melted. Pour over vegetables. Bake uncovered in a 375-degree oven for 15 minutes.

*Yields 6 to 8 servings.*

# A Presidential Picnic

Jacob McGavock Dickinson, a cousin to the Jackson family, purchased the Belle Meade house in 1906. A year later, he was elected president of the American Bar Association. Upon hearing of the plans for the secretary of war, William Howard Taft, to visit Nashville in 1908, Dickinson invited him to dine at Belle Meade. Barbecue was a tradition at the plantation as evidenced by one of the largest smokehouses in the state of Tennessee on the Belle Meade property. A huge barbeque was held on the front lawn of the house and nearly three hundred were in attendance. Taft, known for a hearty appetite, must have enjoyed himself. Upon Taft's election as president, Dickinson was appointed secretary of war. It undoubtedly pays to have connections. This picnic has been recreated at the plantation several times for various occasions.

# HERBED GREEN BEANS

| | |
|---|---|
| 2 pounds young, tender green beans, snapped | 1 1/4 teaspoons chopped mint |
| 6 tablespoons olive oil | 2 1/2 tablespoons minced parsley |
| 2 cloves garlic, minced | 2 1/2 cups tomatoes, peeled and chopped |
| 2 1/2 tablespoons chopped fresh basil | salt to taste |

Assemble all ingredients and utensils. Soak beans for 15 minutes in water to cover; drain. Combine remaining ingredients; sauté for 5 minutes. Blanch beans in boiling salted water for 1 minute; drain and plunge into cold water. Drain. Add to sautéed herbs and cook for 15 minutes.

*Yield: 8 servings*

# SWEET POTATOES AND APRICOTS

| | | | |
|---|---|---|---|
| 1 | 1-pound can whole sweet potatoes, halved lengthwise | 1/4 | teaspoon cinnamon |
| 1 1/4 | cups brown sugar | 1 | teaspoon grated orange rind |
| 1 1/2 | tablespoons cornstarch | 1 | 1-pound can apricot halves |
| 1/4 | teaspoon salt | 2 | tablespoons butter |
| | | 1/2 | cup pecan halves |

Assemble all ingredients and utensils. Place sweet potatoes in greased 2-quart baking dish. In saucepan, combine brown sugar, cornstarch, salt, cinnamon, and orange rind. Drain apricots, reserving syrup. Stir 1 cup apricot syrup into cornstarch mixture. Cook and stir over medium heat until boiling. Boil 2 minutes. Add apricots, butter, and pecan halves. Pour over potatoes. Bake in a 375-degree oven for 25 minutes.

*Yields 6 servings.*

# CHICKEN POTPIE

| | | | | |
|---|---|---|---|---|
| 2 | carrots | 1/2 | teaspoon dried thyme |
| 2 | ribs celery | 1/8 | teaspoon ground sage |
| 1 | medium potato, peeled and chopped | 1 | teaspoon salt |
| 1 | medium onion, chopped | 1/2 | teaspoon pepper |
| 6 | tablespoons (3/4 stick) butter | 2 | refrigerated 9-inch pie pastries |
| 6 | tablespoons all-purpose flour | 4 | cups chopped cooked chicken |
| 2 1/2 | cups chicken broth | 1 to 2 | tablespoons shredded Cheddar cheese |
| 1 1/2 | cups half-and-half | | |

Assemble all ingredients and utensils. Cut the carrots and celery into 1/4-inch slices. Combine with the potato and onion in a small amount of water in a saucepan. Cook, covered for 5 minutes or until tender-crisp; drain. Melt the butter in a saucepan and stir in the flour. Cook until golden brown, stirring constantly. Add the chicken broth and half-and-half gradually. Stir in the thyme, sage, salt, and pepper. Cook for 5 minutes or until thickened and smooth, stirring constantly.

Line a 2-quart baking dish with one of the pie pastries. Sprinkle the chicken into the prepared dish and spread the vegetables over the chicken. Pour the sauce over the top and mix gently. Sprinkle with the cheese.

Top with the remaining pie pastry; seal and flute the edges and cut vents in the top. Bake at 425-degrees for 25 to 30 minutes or until the crust is golden brown.

*Yields 8 servings.*

# TENDERLOIN POT ROAST

| | |
|---|---|
| 1 | 3-pound beef tenderloin roast |
| 1/4 | cup soy sauce |
| 1 | tablespoon Worcestershire sauce |
| 1 | clove garlic, crushed |
| 2 | onions, quartered |
| 1/2 | pound fresh mushrooms sliced |
| 3 | ribs celery, sliced |
| 3 | carrots, peeled and sliced |
| 1/4 | cup water |
| 4 to 6 | small new potatoes, halved |

Assemble all ingredients and utensils.

Preheat the oven to 325 degrees. Place tenderloin in a bowl. In a separate small bowl, combine soy sauce, Worcestershire, and garlic. Pour the mixture over tenderloin and marinate overnight. In a roasting pan, place marinated tenderloin, onions, mushrooms, celery, and carrots. Cover. Bake for 2 hours and 30 minutes. Add the potatoes and water, and bake for another 30 minutes.

*Yields 12 servings of 3 ounces of tenderloin, plus divided vegetables, per person.*

# REUBEN PIE

| | |
|---|---|
| 1 | 9-inch deep-dish pie shell, unbaked |
| 1 | tablespoon caraway seeds |
| 1/2 | pound deli corned beef, shredded |
| 1 | tablespoon Dijon mustard |
| 1/4 | cup Thousand Island dressing |
| 3/4 | cup sauerkraut, drained |
| 1 1/2 | cups grated Gruyère cheese |
| 3 | eggs, beaten |
| 1 | cup half-and-half |
| 1 | tablespoon grated onion |
| 1/4 | teaspoon dry mustard |
| 1/2 | teaspoon salt |
| | Kosher dill spears for garnish |

Assemble all ingredients and utensils.

Preheat the oven to 425 degrees. Sprinkle and press caraway seeds into the unbaked pie crust. With a fork, prick the crust and bake for 7 minutes. Remove the crust and reduce the oven temperature to 350°. Layer corned beef on top of the crust. Combine mustard and dressing and spread over beef, then layer sauerkraut and cheese. Mix eggs, half-and-half, onion, dry mustard, and salt; pour evenly over the pie. Bake for 40 to 45 minutes. Remove from the oven and allow to set for 5 minutes. Garnish each plate with a dill spear.

*Yields 6 servings.*

# APPLE BAKED PORK ROAST

| | | | |
|---|---|---|---|
| 1 | (4- to 5-pound) rolled bone-less pork loin roast | 1/2 | teaspoon garlic powder |
| 1 | teaspoon dried whole rosemary, crushed | 3 | tablespoons apple jelly |
| | | 2 | tablespoons honey mustard |
| 1/2 | teaspoon salt | 2 | tablespoons apple juice |
| | | 1 | tablespoon brown sugar |

Assemble all ingredients and utensils. Place roast, fat side up, on a rack in a shallow roasting pan. Rub roast with rosemary; sprinkle with salt and garlic powder. Insert meat thermometer, making sure it does not touch fat. Bake in a 325-degree oven, 1 hour and 45 minutes. Remove roast from oven; leave oven on.

Combine apple jelly, honey mustard, apple juice, and brown sugar, stirring well. Brush roast with jelly mixture. Continue to bake at 325 degrees for 15 to 30 minutes, or until thermometer registers 160-degrees.

*Yields 8 to 10 servings.*

# WILD RICE CASSEROLE

1 1/2  cups boiling water
1    4.4-ounce package Wild Rice and Herbs
1    10 3/4-ounce can chicken with rice soup
1    small can mushrooms and liquid
1/2   cup water
1    teaspoon salt
1    bay leaf
1/4   teaspoon each: celery salt, garlic powder,
     pepper, onion salt and paprika
3    tablespoons chopped onion
3    tablespoons vegetable oil
3/4   pound lean ground beef

Assemble all ingredients and utensils.

Pour boiling water over rice and herbs. Simmer covered, 15 minutes. Place rice in a 2-quart casserole. Add soup, mushrooms, and liquid, water and seasonings. Mix and set aside. Sauté onion in oil until transparent and add to casserole. Brown meat in frying pan and add to rice mixture. Chill. When ready to bake, cover in a 325-degree oven for 30 to 35 minutes.

*Yields 4 to 6 servings.*

# BEEF CASSEROLE

1    pound ground round, browned and drained
4    small baking potatoes, peeled and sliced
1    2-ounce can sliced mushrooms, undrained
1    8 1/2-ounce can green peas, undrained
3    carrots, sliced
3    stalks celery, sliced
1/2   cup chopped onion
1    teaspoon salt
1    10 3/4-ounce can tomato soup

Assemble all ingredients and utensils. In a 3-quart greased casserole place all ingredients in layers. Pour the soup over top. Cover and bake in a 350-degree oven for 50 to 60 minutes. Optional: top with 1 cup grated cheddar cheese and brown for the last 10 minutes.

*Yields 6 to 8 servings.*

## OLD FASHIONED BOILED CUSTARD

1/2 gallon milk
9 eggs
1 1/2 cups sugar
1/8 teaspoon salt
2 tablespoons corn starch
4 teaspoons vanilla extract

Assemble all ingredients and utensils. In top of a double boiler heat milk. In a large bowl beat eggs; add sugar, salt, and cornstarch. Mix well. Pour a small amount of the hot mixture, 1/4 cup, over the egg mixture. Stir well. Combine with remaining milk and cook until thick. Flavor with vanilla and refrigerate until thoroughly chilled.

*Yields 10 to 12 servings.*

## LEMON SOUR CREAM PIE

1 cup sugar
3 tablespoons cornstarch
1 cup milk
3 egg yolks, beaten
1/4 cup lemon juice
1 tab grated lemon rind
1/4 cup butter
1 cup sour cream
1 baked 9-inch pie crust
1 cup heavy cream, whipped, sweetened with 2 tablespoons sugar

Assemble all ingredients and utensils. In a heavy saucepan, combine sugar and cornstarch. Add milk, egg yolks, lemon juice, rind, and butter. Cook over medium-high heat until thick. Remove from the heat. Cool thoroughly. Add sour cream and pour into a baked pie shell. Top with whipped cream. Refrigerate for several hours.

*Yields 6 to 8 servings.*

# BIG ORANGE COOKIES

1 pound candy orange sliced, finely cut
2 1/2-ounce cans flaked coconut
1 teaspoon orange flavoring
1 teaspoon vanilla
2 cans sweetened condensed milk
1 cup pecans, finely chopped confectioners' sugar, sifted

Assemble all ingredients and utensils.

In large mixing bowl, combine all ingredients except for confectioner's sugar, and mix well. Spread mixture in a lightly oiled 10 x 15 x 1/2 inch baking pan and bake in a 275 degree oven for 30 minutes. Remove from oven; while still hot, spoon mixture into a medium-sized bowl of sifted confectioner's sugar. (Not too much of the hot mixture at a time.) Roll into balls the size of small walnuts and place on waxed paper to cool. Store in airtight container.

*Yields 6 dozen cookies.*

# OLD RECIPE BANANA PUDDING

| | | | |
|---|---|---|---|
| 1 1/2 | cups sugar | 1 | tablespoon vanilla extract |
| 1/4 | cup all-purpose flour | 1 | 16-ounce package |
| 1/2 | teaspoon salt | | vanilla wafers |
| 4 | cups whole milk | 4 | bananas, sliced |
| 6 | egg yolks, beaten | 1/2 | cup sugar |

Assemble all ingredients and utensils. In top of double boiler, combine 1 1/2 cups sugar, flour, and salt. Stir in milk. Cook over boiling water until thickened, stirring constantly. Remove from heat. Stir a small amount of hot mixture, 1/4 cup, into egg yolks. Return egg yolk mixture to hot mixture in double boiler. Cook over boiling water, stirring constantly for about 5 minutes. Remove from heat. Stir in vanilla.

Layer wafers, bananas, and custard alternately in a 2-quart baking dish. Beat egg whites until stiff and peaks form. Beat in 1/2 cup sugar. Spread over top of pudding. Bake in a 350-degree oven for 8 to 10 minutes or until lightly browned.

*Yields 10 to 12 servings.*

## Collecting Silver

General Harding was known for breeding many types of livestock at the Belle Meade farm. Herds of Jersey cattle, Cashmere goats, Shetland ponies, and thoroughbred horses were all bred on the 5000-acre estate. Harding showed his stock at local fairs hosted by agricultural and mechanical associations in the 1850s and 1860s. Silver premiums were awarded to the winners, and Harding amassed a vast collection of silver. Other members of the family also added to the collection with their winning trophies for various other activities. A visitor following the Civil war said that Harding must have owned more silver and plate than any other man in the South. Belle Meade continues that tradition today with the collecting of silver from family members, as well as yearly trophies from the Iroquois Steeplechase engraved with the winner's names.

# CARROT CAKE

| | |
|---|---|
| 2 cups sifted all-purpose flour | 2 cups finely grated carrots |
| 2 teaspoons baking powder | 1 8 1/2-ounce can crushed |
| 1 1/2 teaspoons baking soda |    pineapple, drained |
| 1 1/2 teaspoons salt | 1/2 cup chopped pecans, optional |
| 2 teaspoons cinnamon | 1 3 1/2-ounce can flaked |
| 2 cups sugar |    coconut cream |
| 1 1/2 cups salad oil |    cheese frosting |
| 4 eggs | |

Assemble ingredients and utensils. In a large bowl or mixer, sift together flour, baking powder, baking soda, salt, and cinnamon. Gradually mix in sugar, oil, and eggs; beat well. Add carrots, pineapple, nuts, and coconut; blend thoroughly. Distribute batter evenly among three, 9-inch greased and floured round cake pans. Bake in a 350-degree oven for 35-40 minutes. Cool about 10 minutes in the pans. Turn out on wire racks and cool thoroughly. Fill layers and frost top and sides of cake with cream cheese frosting.

*Yields 15-18 servings.*

# STRAWBERRY JAM CAKE

| | |
|---|---|
| 2 cups flour | 3 eggs, beaten |
| 1 cup light brown sugar | 1 teaspoon baking soda |
| 1 teaspoon cinnamon | 1/2 cup buttermilk |
| 1/2 teaspoon cloves | 1 1/2 cups strawberry jam |
| 1/4 teaspoon salt | 1/2 cup chopped pecans |
| 1 cup butter, softened | |

Assemble ingredients and utensils. Sift dry ingredients in a large bowl. Gradually cream in butter and add eggs. Mix well. Dissolve baking soda in buttermilk and mix into creamed mixture. Continue beating for 3 minutes. Add jam and pecans, mix well. Grease and flour a 9x13-inch pan. Pour batter into pan and bake in a 350-degree oven for 55-60 minutes. Cool. Ice with easy caramel icing.

*Yields 24 servings.*

## CARMEL ICING

| | |
|---|---|
| 1 1/2 cups brown sugar | 1 teaspoon vanilla |
| 1/4 cup whole milk | small amount of |
| 2 tablespoons butter | whipping cream, optional |

In a large saucepan, combine sugar, milk, and butter; bring to a boil and boil for 3 minutes, stirring constantly. Remove from heat; add vanilla. Cool to lukewarm. Beat until creamy and thick enough to spread. You may add a little cream if necessary to make it easier to spread.

*Yields 1 1/2 to 2 cups.*

# APRICOT DAINTIES

1   6-ounce package dried apricots, chopped
1   cup sugar
3   tablespoons orange juice
1   cup finely chopped pecans
    Confectioner's sugar

Assemble all ingredients and utensils.

In the top of a double boiler, cook apricots, sugar, and orange juice until sugar dissolves and apricots soften. Cool the mixture and shape fruit into balls with a pinch of pecans in the center of each. Roll balls in remaining pecans, then in confectioners' sugar.

*Yields approximately 50 balls.*

# PECAN FUDGE PIE

1/2   cup butter
4   1-ounce squares unsweetened chocolate
4   eggs, lightly beaten
3   tablespoons light corn syrup
1 1/2   cups sugar
1/4   teaspoon salt
1   teaspoon vanilla extract
1   cup chopped pecans
1   9-inch pastry shell

Assemble all ingredients and utensils. In top of double boiler or in a saucepan over low heat, melt butter and chocolate. Combine in a bowl beaten eggs, syrup, sugar, salt, vanilla, and pecans. Mix well. Add chocolate mixture and mix thoroughly. Pour filling into the pie shell. Bake in a 350-degree oven for 30 to 35 minutes until filling is set, but soft inside.

*Yields 6 to 8 servings.*

# TENNESSEE BLACKBERRY COBBLER

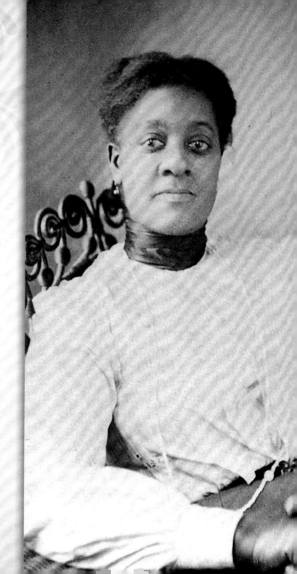

        4   cups fresh blackberries or
       24   ounces frozen blackberries, thawed and drained
    1 1/2   cups sugar
        3   tablespoons all-purpose flour
        1   tablespoon fresh lemon juice
        3   tablespoons butter

Assemble all ingredients and utensils. Fruit mixture: toss berries, sugar and flour, place in an ungreased 9-inch baking dish. Sprinkle berries with lemon juice and dot with butter, set aside.

        2   cups all-purpose flour
      1/4   teaspoon salt
        1   tablespoon baking powder
        1   cup heavy cream, whipped
        1   tablespoon sugar

Assemble all ingredients and utensils. In a large bowl mix flour, salt, and baking powder. Gently fold cream into flour mixture. Place dough on floured board and knead for one minute. Roll dough to 1/2 inch thickness. Cut into lattice strips or place the entire sheet of dough on cobbler. Sprinkle with sugar. Bake in a 400-degree oven for 10 to 12 minutes, then reduce heat to 325 degrees and bake another 20 minutes, or until golden brown.

*Yields 8 servings.*

# ROSY RHUBARB COBBLER

|       |                         |
|------:|-------------------------|
| 4     | cups rhubarb, diced     |
| 1     | cup sugar               |
| 3     | tablespoons butter      |
| 1 1/2 | cups all-purpose flour  |
| 1/4   | teaspoon salt           |
| 3     | teaspoons baking powder |
| 1     | cup sugar               |
| 1/4   | cup shortening          |
| 1     | egg, beaten             |
| 1/2   | cup milk                |

Assemble all ingredients and utensils.

Place rhubarb in greased 8 x 12 inch baking dish. Sprinkle with the 1 cup sugar and dot with butter. Heat in a 350-degree oven while mixing batter. Sift remaining dry ingredients and cut in shortening until mixture resembles coarse crumbs. Add beaten egg which has been mixed with the milk. Pour batter over hot rhubarb. Bake in a 350-degree oven for 35 minutes, until browned. Serve warm.

*Yields 6 servings.*

THE WINERY AT
BELLE MEADE PLANTATION

# THE WINERY AT BELLE MEADE PLANTATION

Hospitality was always a cornerstone of Belle Meade. Spirits and wine were part of the culture of entertainment on the plantation. Ledgers show extensive purchases of brandy, whiskey, and fine wines from around the world in the 1880s and 90s. They also show ongoing purchases of empty wine bottles in the fall, presumably to make their own wines. Today, muscadine vines grow profusely along fence rows. This was most likely the grape of choice for homemade wine and was a Middle Tennessee tradition.

That tradition lives on today as The Winery at Belle Meade Plantation continues the centuries old art form of turning grapes into wine for their guests. Like most traditions on the plantation, the fine wines open the window to the past and let one sip a tradition that goes back two hundred years!

THE WINERY AT
BELLE MEADE PLANTATION
NASHVILLE, TENNESSEE

IROQUOIS RED

CABERNET SAUVIGNON

IROQUOIS WAS THE FIRST AMERICAN BRED STALLION TO WIN ALL THREE PREMIER ENGLISH EVENTS. (THE ENGLISH DERBY, PRINCE OF WALES, AND THE ST. LEDGER STAKES)

# RUM CREAM PIE

|   |   |
|---|---|
| 1 | 9-inch graham cracker pie shell |
| 6 | egg yolks |
| 1 | cup sugar |
| 1 | envelope unflavored gelatin |
| 1/2 | cup cold water |
| 1 | pint heavy cream |
| 1/2 | cup rum |
|   | bittersweet chocolate shavings |
|   | whipped cream for garnish, optional |

Assemble all ingredients and utensils. Beat egg yolks until light and add 1 cup sugar. Soak gelatin in 1/2 cup cold water. Put the gelatin and water over low heat and bring to a boil. Pour this over the sugar-egg mixture, stirring briskly. Whip cream until stiff; fold into the egg mixture. Add rum. Cool until the mixture begins to set and then pour into the pie shell. Chill until firm; sprinkle top of pie generously with chocolate shavings and garnish with whipped cream, if desired.

*Yields 6 to 8 servings.*

# BELLE MEADE
## PLANTATION

For General Information, Brochure, or Visit Planning

615.356.0501 or 800.270.3991

info@bellemeadeplantation.com

### BELLE MEADE PLANTATION
#### QUEEN OF TENNESSEE PLANTATIONS

5025 Harding Pike, Nashville, Tennessee, 37205

www.bellemeadeplantation.com

# DESCRIPTION OF PHOTOGRAPHS

*Unless otherwise noted, photographic images are provided courtesy of Belle Meade Plantation Permanent Collection. Many images contributed by Mary Lawson Photography are identified by MLP at the end of each description below. In addition, her supplemental artistic images appear on pages 38, 41, 48, 52, 54, 57, 60, 63, 69, 70, 73, 78, 80, 85, 94, 104, 108, 118, and 120 and are not noted). Special thanks to Caroline Allison Photography for the image on pages 8-9.*

# RECIPE INDEX